Dinosaurs
and their living relatives

Contents

2

Preface

Could dinosaurs – animals that died out 65 million years ago – be closely related to crocodiles and birds alive today? After reading this book, you should be able to decide for yourself.

The book takes a completely new approach to the study of dinosaurs. It sets out to discover how dinosaurs are related to other animals – both living and extinct. It begins by explaining a simple method for working out the relationships between animals. Then, using many photographs and diagrams, it applies this method to the dinosaurs. The book ends with a unique series of new full-colour illustrations of many of the Natural History Museum's most famous dinosaurs – as they may have appeared when they were alive.

Dinosaurs and their living relatives has been produced in conjunction with a permanent new exhibition of the same name that opened at the Natural History Museum in May 1979. This exhibition, which was planned with the guidance of Museum experts, is the third in the Museum's major new exhibition programme.

Preparing the exhibition and its companion book has involved the effort and imagination of a great many people, both within the Museum and outside, and I should like to take this opportunity of thanking everyone concerned.

R. H. Hedley, Director
British Museum (Natural History)
August 1979

Dinosaurs from left to right: *Iguanodon, Altispinax, Polacanthus, Hypsilophodon* (two).

On the next page, from left to right: *Camptosaurus* (two), *Stegosaurus, Brachiosaurus, Apatosaurus.*

Could dinosaurs – animals that died out 65 million years ago – be closely related to crocodiles and birds alive today?

Read on and see how to work out relationships. Then you can decide for yourself.

Chapter 1

Working out relationships

How can we make sense of
so many living things?

We can group similar living things together.

For instance . . . we could pick out all the ones that have six legs.

Living things that are similar in one way are often similar in other ways too. You may know other similarities that these animals share.

Look back at the picture on the left and try to pick out some other groups. What do the members of your groups have in common?

It's not always as easy as it looks.

Sometimes it is difficult to decide which group a living thing belongs to. For example, some of the living things in these six pictures belong to the group we call animals.

Which do you think they are?

You can find out whether you are right on page 71.

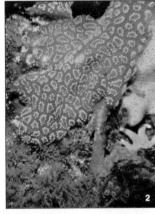

We can make sense of living things by grouping similar ones together. But, how can we know whether our groups consist of **closely related** living things?

7

Are all living things related?

If we assume that life on Earth appeared only once (probably about 4000 million years ago), then all living things must be related to each other. They must all share a **common ancestor.**

We think that they evolved from this single common ancestor by a process of very gradual change over an extremely long time.

**DICE
DINE
DIRE
DARE
CARE
CARD**

Small changes over a long time may produce living things that are quite different from their ancestors – over many generations, species change and become new species. This process of change is called evolution.

The diagrams show three ways in which new species could arise.

One species could change gradually and become a new species.

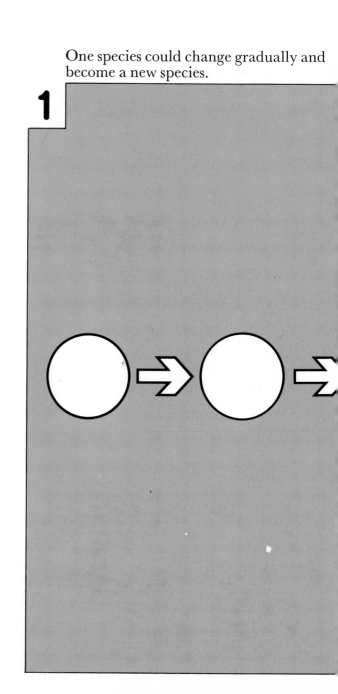

One species could split into two new species.

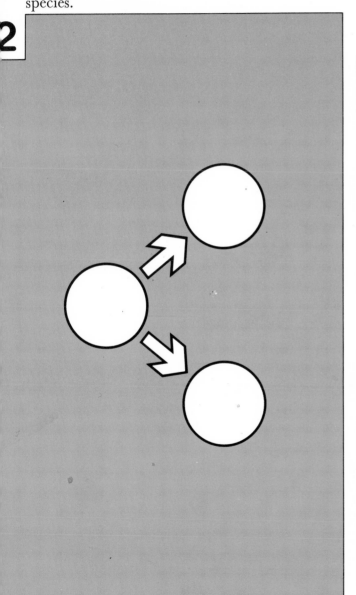

One species could split into many new species at **exactly** the same time.

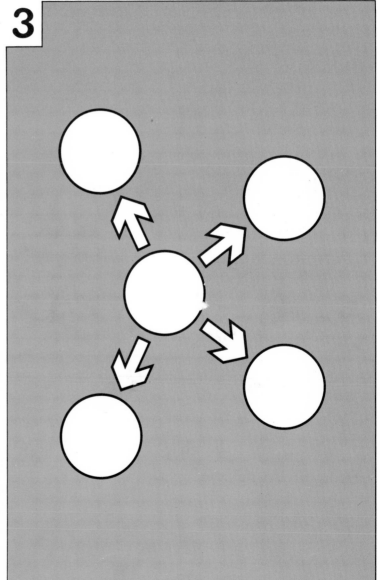

We cannot go back in time, so we can never **know** how species have arisen. But, if we assume that new species arise when one species splits into two (**2** on the previous page), we can suggest relationships that can be tested.

Let's consider three species . . .

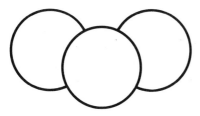

To work out how they are related, we assume two things.
First, we assume that they arose as in **2**.
Second, we assume that none of the species is the ancestor of the others.

So the three species must be related as shown on the right – two of them share a more recent common ancestor than either shares with the third.

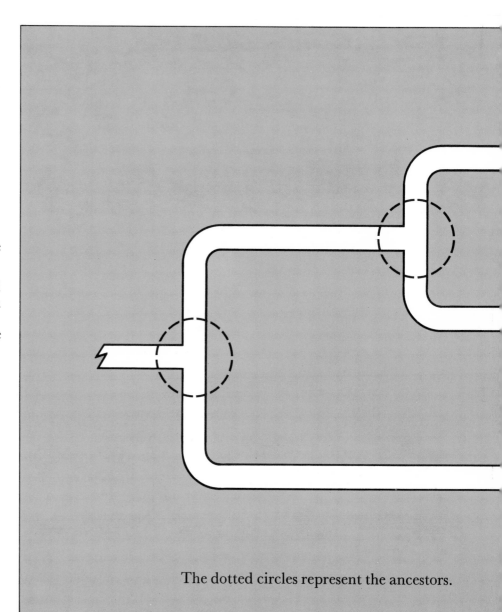

The dotted circles represent the ancestors.

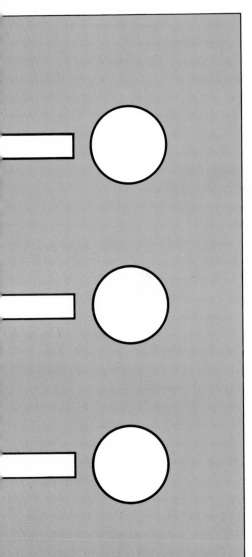

Let's call the species A, B and C...

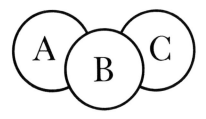

They must be related in one of the three ways shown on the right. (You can work out for yourself that these are the only alternatives.)

We decide which of the ways is most likely, and then we test our ideas to see if we are right.

This is how we work out relationships.

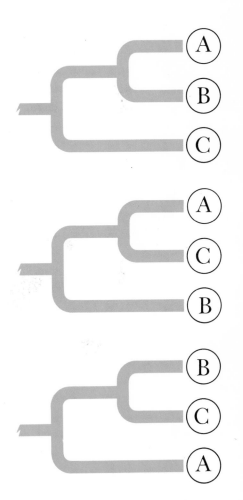

You can find out how we test our ideas in the next chapter.

Chapter 2
Testing our ideas

Plants and animals inherit characteristics from their ancestors. Characteristics that are similar because they are inherited from a common ancestor are called **homologies**. If animals share a homology, we assume they share a common ancestor that also had the homology.

We use homologies to test our ideas about the way animals are related.

These four animals share a number of homologies. Look carefully at the pictures and see if you can spot one of them. The bones have been coloured to help you.

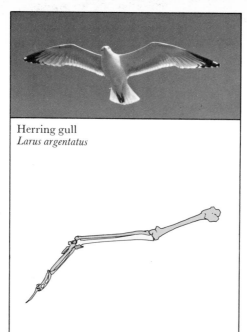

Herring gull
Larus argentatus

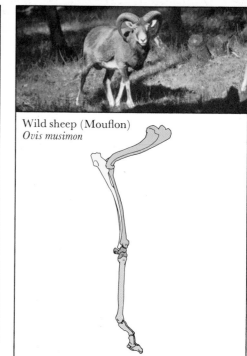

Wild sheep (Mouflon)
Ovis musimon

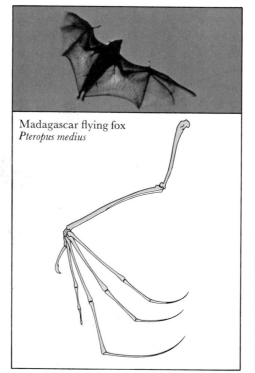

Madagascar flying fox
Pteropus medius

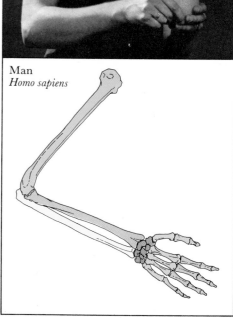

Man
Homo sapiens

Did you notice that all the animals have the same arrangement of bones in their front limbs? This is one of the homologies that they share.

Consider these three animals – a sheep, a herring gull and a salmon. Two of them must be more closely related to each other than either is to the third. So they must be related in one of the three ways shown on the right.

Which do you think is most likely? Why?

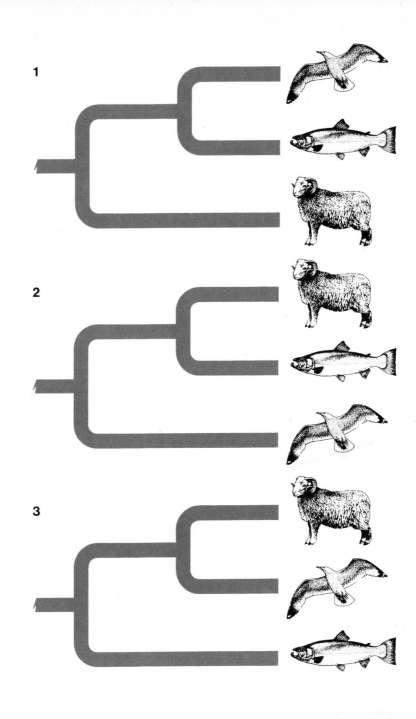

1

2

3

The sheep and herring gull share a homology (the arrangement of bones in the forelimbs) not shared by the salmon. So we think that, of the three, the sheep and herring gull are the two most closely related. So **3** is correct.

More animals . . . more homologies

When we consider more animals, we need to find more homologies.

So how do we fit in this fruit bat?

We know that the fruit bat has the same arrangement of bones in its front limb as the herring gull and sheep, but to which of the two is it most closely related?

To decide, we need to find another homology – one shared by just two of the three animals.
Can you think of one?

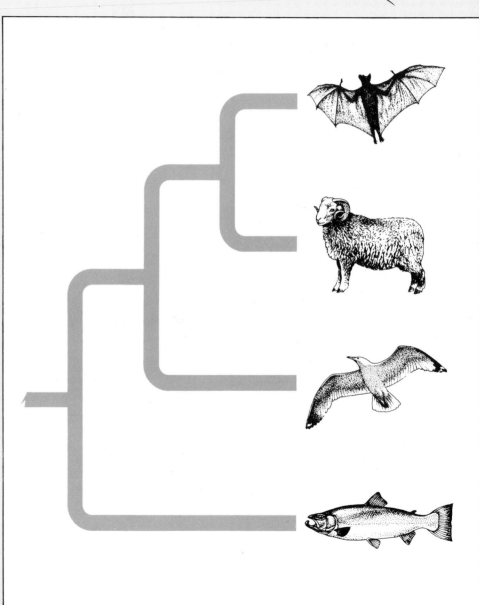

The sheep and fruit bat both have hair, but the herring gull does not. So we think that the sheep and bat are the two most closely related.

These relationships are represented on this branching diagram.

Recognizing homologies can be difficult

Some structures are similar because they have a similar function and **not** because they were inherited from a common ancestor. Here are two examples.

Tortoises and armadillos both have shells to protect them, but their shells are made of different structures – the ribs form part of the shell in tortoises, but not in armadillos. We think that their shells evolved independently, so this is not a homology.

Spur-thighed tortoise *Testudo graeca*

Hairy armadillo *Chaetophractus nationi*

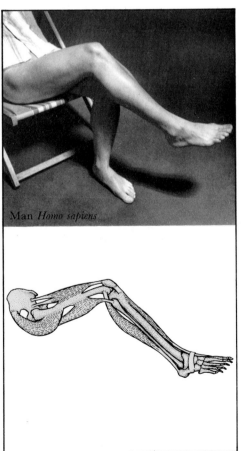

Man *Homo sapiens*

Desert locust *Schistocerca gregaria*

We have legs and so do locusts, but the structures are really quite different. Our muscles are attached to the outside of our skeletons. Locusts have their muscles inside their skeletons.

In both these examples, the similarities are not homologies. So they cannot be used for working out relationships.

Can you recognize homologies?

Two of these animals have structures that are homologous to a bird's wing.
Which do you think they are?

The bat and whale have front limbs that are homologous to a bird's wing – the arrangement of the bones is the same.

The fly's wing has no bones at all, and the bony rays in the flying fish wing are not homologous to the bones in a bird's wing.

More homologies . . .
more closely related

We assume that the more homologies two animals share, the more recent is their common ancestor, and therefore the more closely they are related.

Animals that are very closely related share homologies not shared by any other animal – they share **unique homologies.**

For example, all whalebone whales (e.g. blue whales, fin whales, humpback whales) have baleen plates to sieve their food. No other animal shares this homology – it is unique to whalebone whales.

Southern right whale, a whalebone whale

Skull of whalebone whale, showing baleen

Always choose the simplest explanation

Do you think the dolphin is more closely related to the salmon or to the squirrel?

Study these pictures and try to decide. To help you, the information is summarized below.

Dolphin

Teeth A dolphin has two sets of teeth during its life.

Heart A dolphin's heart has four chambers.

Shape A dolphin's body is stream-lined with short flippers.

Mammary glands Dolphins have mammary glands for feeding their young with milk.

Breathing A dolphin has lungs, and breathes air at the water surface.

Salmon

SUMMARY	Dolphin	Salmon	Squirrel
two sets of teeth	yes	no	yes
four-chambered heart	yes	no	yes
stream-lined shape	yes	yes	no
mammary glands	yes	no	yes
breathes air	yes	no	yes

Teeth A salmon has one set of teeth that grow throughout its life.

Heart A salmon's heart has only two chambers.

Shape A salmon has a stream-lined body with short fins.

Mammary glands Salmon do not have mammary glands.

Breathing A salmon breathes underwater, using gills.

Squirrel

Teeth A squirrel has two sets of teeth during its life.

Heart A squirrel's heart has four chambers.

Shape A squirrel has a short body with four legs and a bushy tail.

Mammary glands A squirrel feeds it young with milk from its mammary glands.

Breathing A squirrel has lungs and breathes air.

The dolphin is a similar shape to the salmon, and they both have 'fins', but the dolphin shares more homologies with the squirrel than with the salmon.

It is simpler to assume that the stream-lined shape evolved twice, than that all the characteristics shared by the dolphin and the squirrel evolved twice. So we assume that the dolphin is more closely related to the squirrel.

When there is conflicting evidence, we always choose the simplest explanation.

Chapter 3

Forming groups

We can represent the way we think animals are related by constructing a branching diagram, which we call a **cladogram.** On a cladogram we can pick out groups consisting of all the animals that share a common ancestor. We call these groups **clades** or true groups.

The diagrams below show a cladogram with four clades. On each diagram, a different clade is coloured.

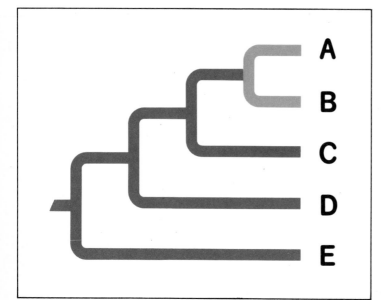

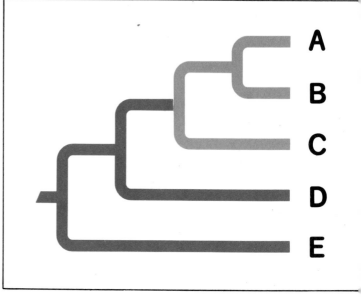

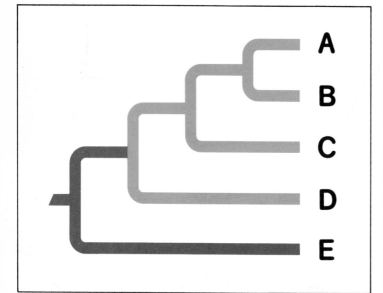

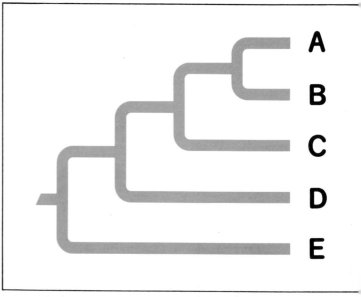

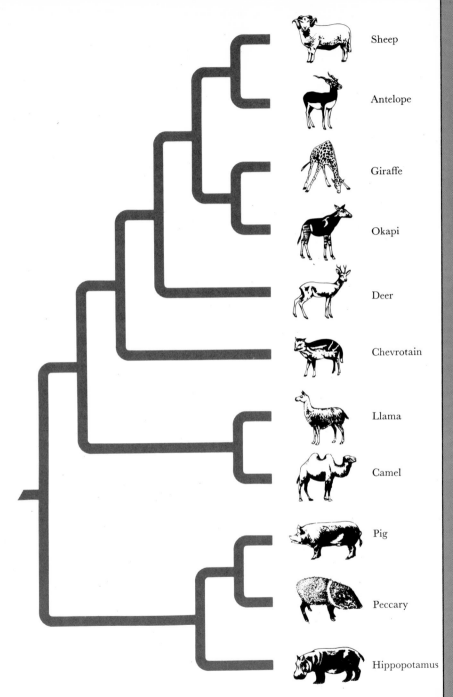

Sheep

Antelope

Giraffe

Okapi

Deer

Chevrotain

Llama

Camel

Pig

Peccary

Hippopotamus

Can you recognize clades?

Look at the cladogram on the left. Do you think the following groups of animals are clades?

Group 1:
sheep, antelope, giraffe, okapi

Group 2:
deer, chevrotain, llama

Group 3:
pig, peccary, hippopotamus.

Answer Groups 1 and 3 are clades because they include all the descendants of a common ancestor.

Not all familiar groups are clades . . .

We often group together animals that share a number of similar characteristics, but these groups are not always clades.

To test whether a group is a clade, we must see if the animals share a homology not shared by any other animal – a unique homology.

Do fishes form a clade?

It has a backbone

It has jaws . . .

It has scales

It has gills

. . . Do any other animals have a backbone?

. . . Do any other animals have jaws?

. . . but crocodiles do too.

. . . but so do tadpoles.

Fishes do not share a unique homology, so the group does not include **all** the descendants of a common ancestor. It is not a clade.

Cladogram of fishes and other backboned animals

What do you notice about the common ancestor of fishes? Is it the ancestor of any other animals? (The common ancestor of fishes is indicated by the circle.)

Most of us are more familiar with traditional groups than with clades. But when we try to reconstruct the history of life, we use clades because then we can test the relationships we suggest.

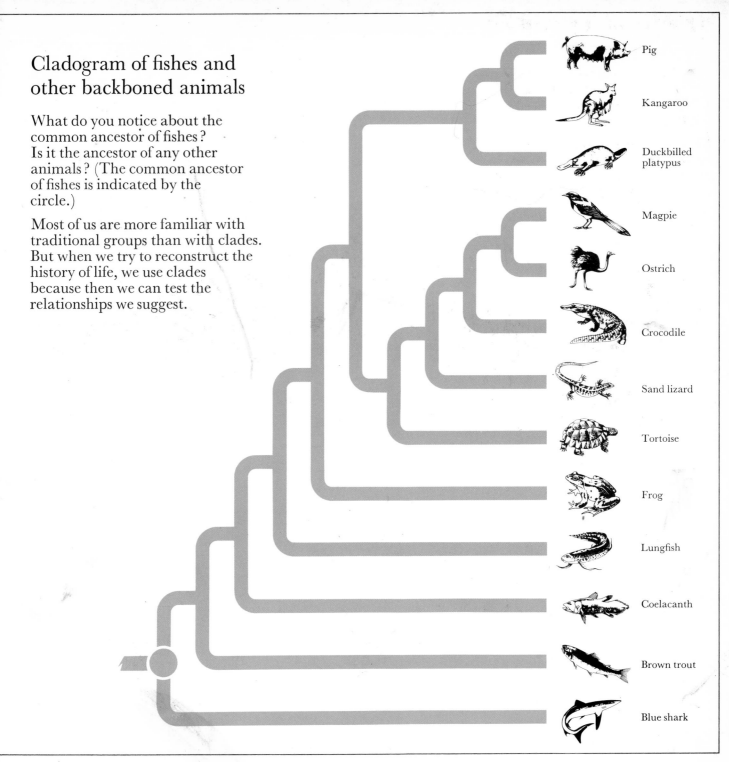

Pig

Kangaroo

Duckbilled platypus

Magpie

Ostrich

Crocodile

Sand lizard

Tortoise

Frog

Lungfish

Coelacanth

Brown trout

Blue shark

Chapter 4

Dinosaurs and their relatives

Now you have seen how to work out relationships, you can work out how dinosaurs are related to other animals.

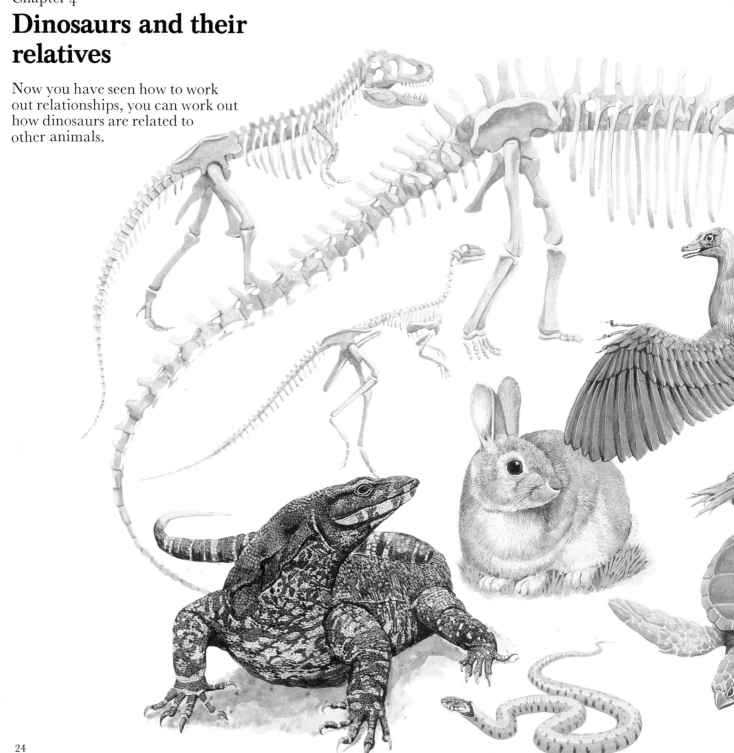

The present is the key to the past

We know more about living animals than about dinosaurs. So we start by working out how living animals are related to each other. Then we fit in the dinosaurs.

25

This cladogram shows how we think birds, crocodiles, lizards and snakes, turtles and tortoises, and mammals are related.

Do these animals form a clade?

To find out, we must see if they share a common ancestor not shared by any other animal. We must look for a homology that they share with each other, but with no other animal – a homology that is unique to them.

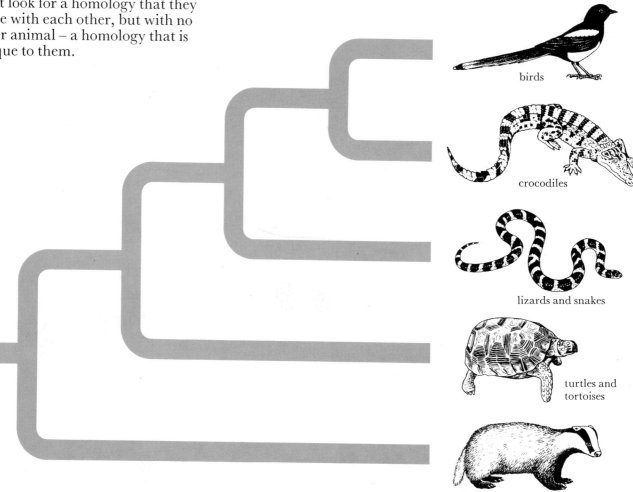

birds

crocodiles

lizards and snakes

turtles and tortoises

mammals

A homology in the egg . . .

When we look at the way these animals develop before they are born, we discover that in all of them the embryo is surrounded by a membrane called the **amnion**. The amnion contains a fluid which keeps the embryo wet and cushions it against bumps and knocks.

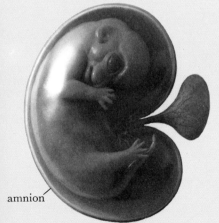

amnion

*turtle embryo

amnion

*lizard embryo

amnion

mammal embryo

amnion

*crocodile embryo

amnion

*bird embryo

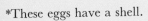

*These eggs have a shell.

. . . but is it unique?

Do any other living animals have an amnion?

All embryos must be protected in some way, but if the eggs are laid in water, as fish, frog and newt eggs are, the water cushions and protects the embryo. These animals do not have an amnion.

Until the time in the history of life when the amnion evolved, all backboned animals had to live near water where they could lay their eggs. Then some animals developed an amnion. This enabled them to live all their lives on land.

Crocodiles, birds, lizards and snakes, turtles and tortoises, and mammals are the only* living animals that have an amnion.

*One other living animal – the tuatara – also has an amnion. We think that the tuatara is closely related to lizards and snakes.

28

Did dinosaurs have an amnion?

We can never know, because we cannot go back in time. But to help us decide, we can use our knowledge of living animals to interpret the fossil remains of extinct ones.

Fossils show us that dinosaurs laid (shelled) eggs on land. All living backboned animals that lay their eggs on land have an amnion. So we assume that dinosaurs had an amnion too.

Protoceratops eggs   scale 1:3

Baby *Protoceratops* hatching (reconstruction based on eggs and embryo skeletons)

Protoceratops is a small dinosaur that is closely related to *Triceratops*. Its fossilized eggs and young have been found in the Gobi Desert in Mongolia.

29

Ichthyosaur

Plesiosaur

Did any other extinct animals have an amnion?

We very rarely know about the soft parts of extinct animals, because usually only their bones and teeth are preserved as fossils.

But we assume that, if extinct animals and living animals share homologies in their bones and teeth, they probably share homologies in their soft parts too.

Ichthyosaurs and **plesiosaurs** share a number of similarities with living animals that have an amnion, so we assume that they had an amnion too.

Some ichthyosaurs have been preserved with their young inside them, so we know that they gave birth to live young. This is further evidence that they had an amnion.

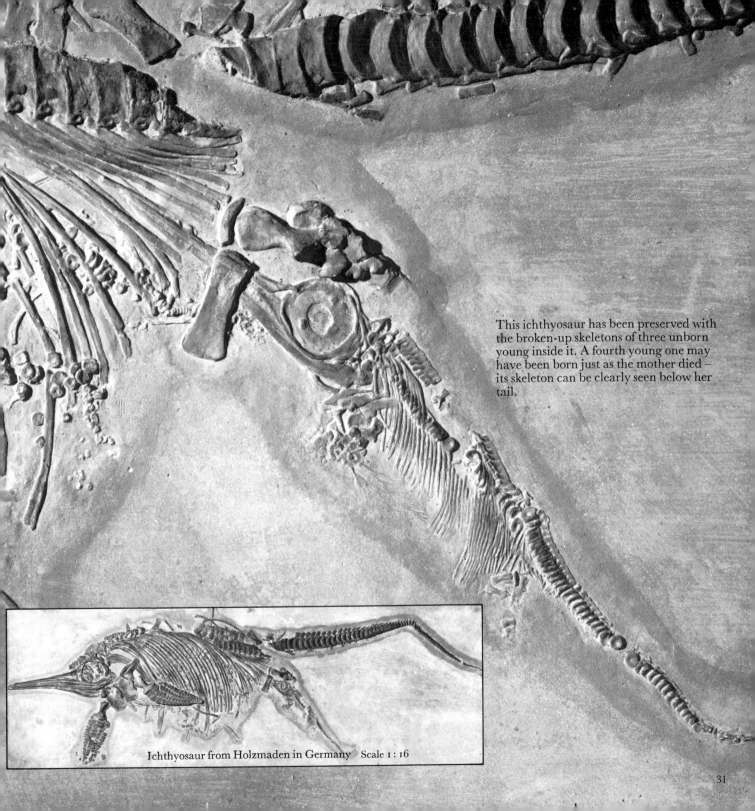

This ichthyosaur has been preserved with the broken-up skeletons of three unborn young inside it. A fourth young one may have been born just as the mother died – its skeleton can be clearly seen below her tail.

Ichthyosaur from Holzmaden in Germany   Scale 1 : 16

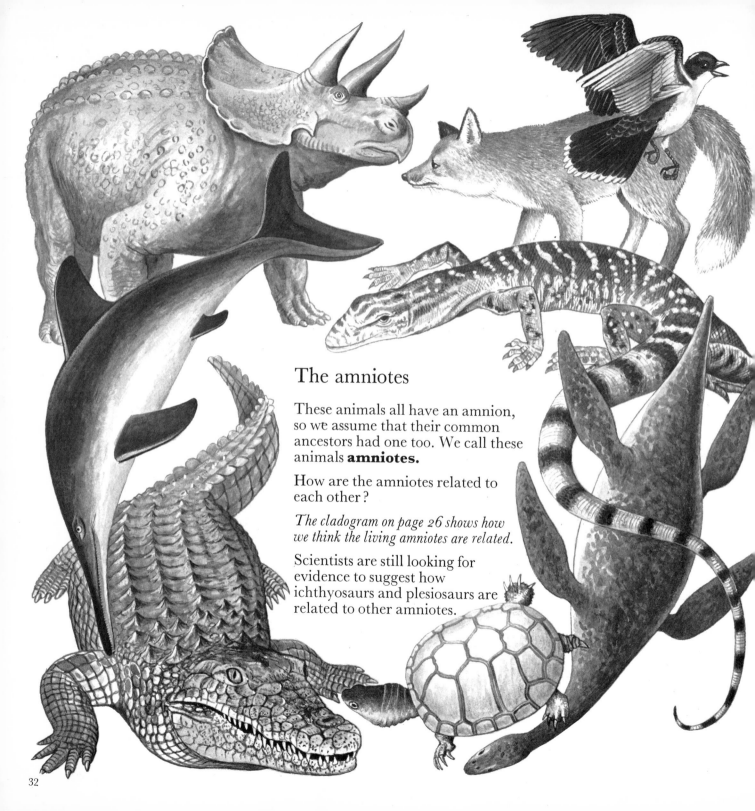

The amniotes

These animals all have an amnion, so we assume that their common ancestors had one too. We call these animals **amniotes.**

How are the amniotes related to each other?

The cladogram on page 26 shows how we think the living amniotes are related.

Scientists are still looking for evidence to suggest how ichthyosaurs and plesiosaurs are related to other amniotes.

How are dinosaurs related to other amniotes?

When we compare the skulls of dinosaurs with those of other amniotes, we find that dinosaurs share a homology with fossil* birds and crocodiles – they have a hole in front of the eye socket.

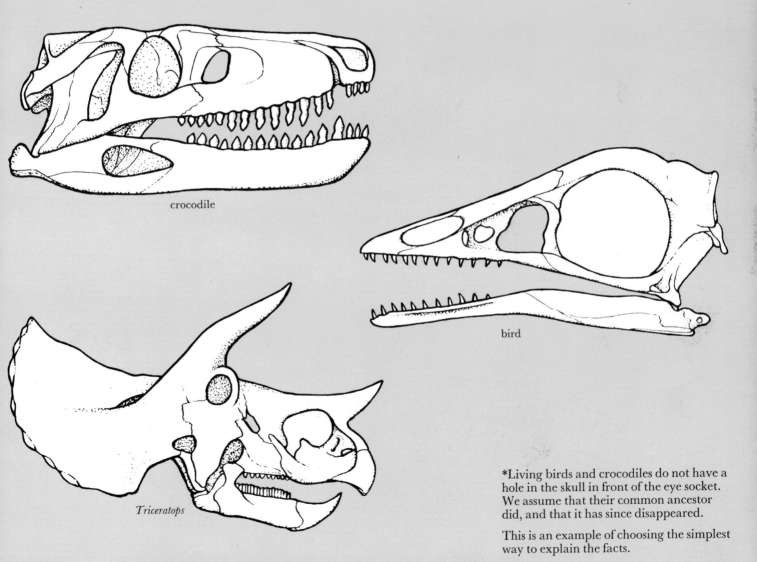

crocodile

bird

Triceratops

*Living birds and crocodiles do not have a hole in the skull in front of the eye socket. We assume that their common ancestor did, and that it has since disappeared.

This is an example of choosing the simplest way to explain the facts.

Does any other animal share this homology?

Two other types of extinct animal – **pterosaurs** and **thecodontians** – also have a hole in the skull in front of the eye socket. This homology is unique to birds, crocodiles, dinosaurs, pterosaurs and thecodontians, so we assume that they shared a common ancestor not shared by any other animal. They form a clade which we call **archosaurs.**

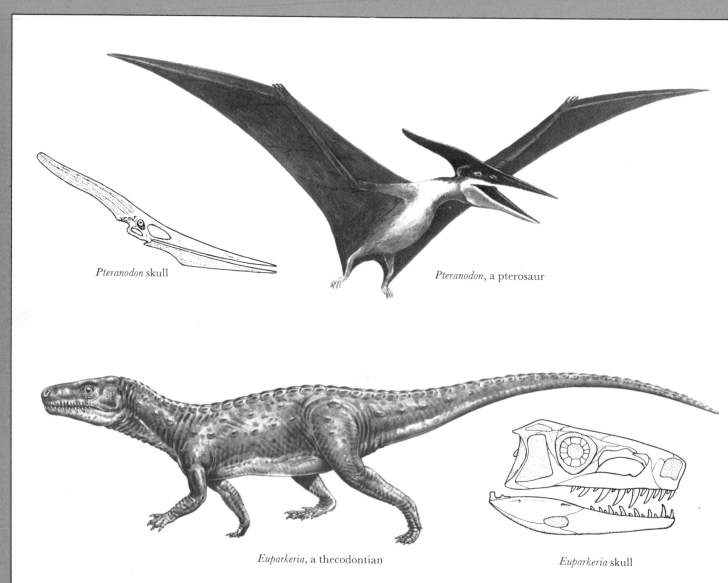

Pteranodon skull

Pteranodon, a pterosaur

Euparkeria, a thecodontian

Euparkeria skull

Cladogram of archosaurs

This cladogram shows how we think the archosaurs are related to each other. (Because we are not sure how fossils are related to each other, we have represented their relationships as broken lines on the cladogram.)

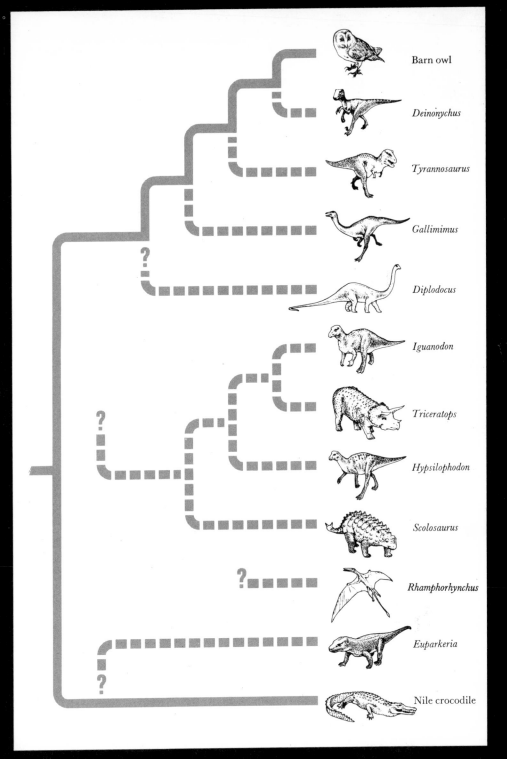

Barn owl

Deinonychus

Tyrannosaurus

Gallimimus

Diplodocus

Iguanodon

Triceratops

Hypsilophodon

Scolosaurus

Rhamphorhynchus

Euparkeria

Nile crocodile

In the next chapter you can see how to test these ideas by looking for homologies.

Chapter 5

Two types of dinosaur

Scolosaurus
Up to 5 metres long

Diplodocus
Up to 26 metres long

Gallimimus
Up to 4.5 metres long

Hypsilophodon
Up to 1.5 metres long

Iguanodon
Up to 9 metres long

Tyrannosaurus
Up to 15 metres long

Triceratops
Up to 11 metres long

When we look at dinosaur skeletons, we find that their hips are of two distinct types.

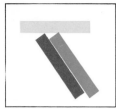

Dinosaurs with hips like this are called **ornithischians.**

Dinosaurs with hips like this are called **saurischians.**

Which of these dinosaurs are ornithischians?
Which are saurischians?
Look carefully at the skeletons, and decide for yourself.

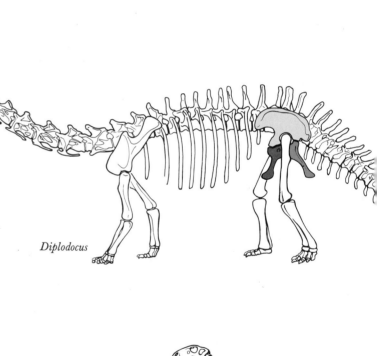

Scolosaurus

Diplodocus

Gallimimus

Hypsilophodon

Iguanodon

Tyrannosaurus

Triceratops

Ornithischian dinosaurs

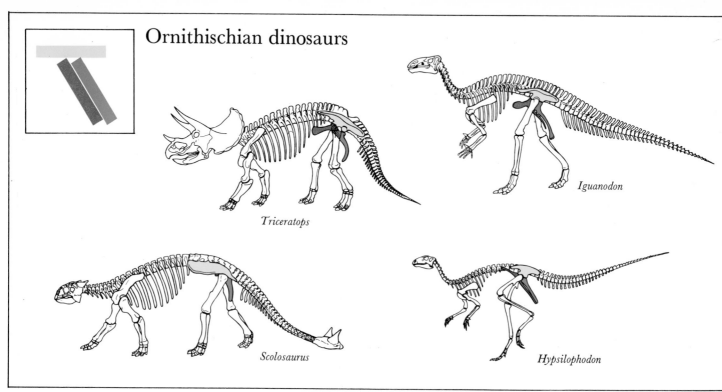

Triceratops

Iguanodon

Scolosaurus

Hypsilophodon

Saurischian dinosaurs

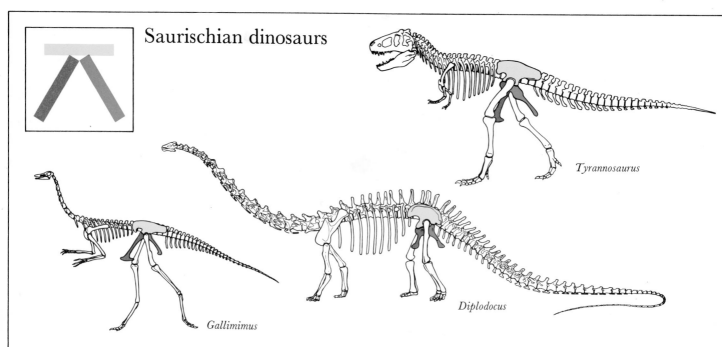

Tyrannosaurus

Gallimimus

Diplodocus

Do saurischian dinosaurs form a clade?

crocodile

Euparkeria, a thecodontian

The saurischian type of hip is not unique to saurischians – crocodiles and thecodontians have similar hips.

At the moment we know of no unique homology that saurischians share, so we cannot call them a clade.

Do ornithischian dinosaurs form a clade?

The ornithischian type of hip is not unique to ornithischians – birds have similar hips.

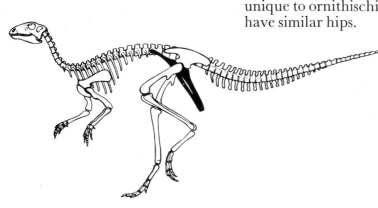

But ornithischians do share a unique homology – they all have a predentary bone in the lower jaw. No other archosaur has a predentary bone.

Because they share a unique homology, we assume that ornithischians share a common ancestor not shared by any other animal. They form a clade.

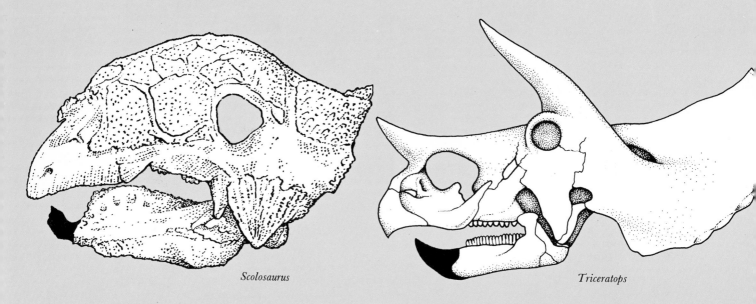

Scolosaurus

Triceratops

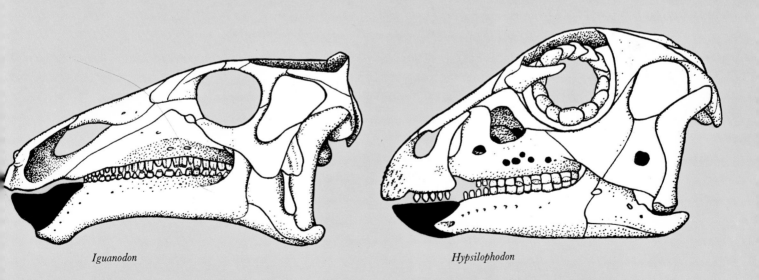

Iguanodon

Hypsilophodon

On the next page, you can look for
homologies to test the relationships
between individual ornithischians.

Ornithischian dinosaurs — how are they related to each other?

There are **fifteen** different ways in which these four ornithischians could be related to each other. (You could work out for yourself the fifteen different cladograms.)

But by making just one decision, we can reduce the possibilities to a more manageable number.

If we can decide which two ornithischians are most closely related, we will reduce the possibilities from fifteen to **three**.

Hypsilophodon

Iguanodon

Scolosaurus

Triceratops

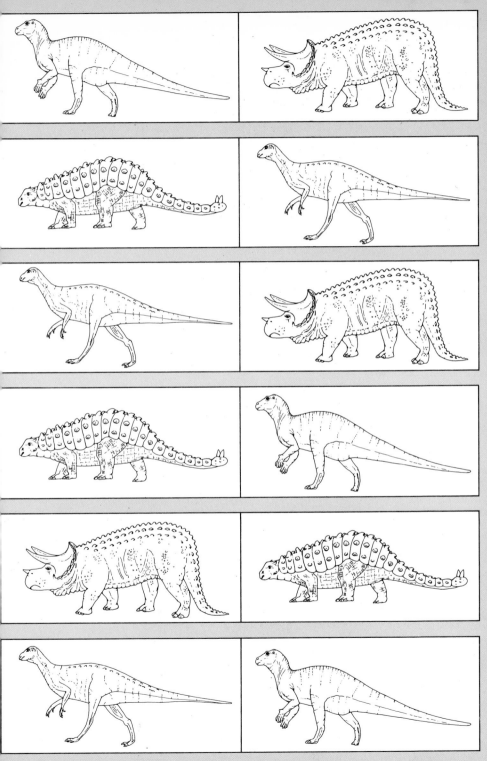

Which two ornithischian dinosaurs are most closely related?

Compare their hips and decide for yourself. Look especially at the shape and length of the pubis – the red bone.

You can check your answer on the next page.

Do you agree that *Iguanodon* and *Triceratops* are the two most closely related?

If this is so, then these three cladograms show the only possible ways the four ornithischians could be related to each other.

To decide which cladogram is most likely to be correct, look again at page 44. Which of the two remaining dinosaurs has a hip most similar to the hips of *Iguanodon* and *Triceratops*?

Look carefully, especially at the shape and length of the pubis – the red bone.

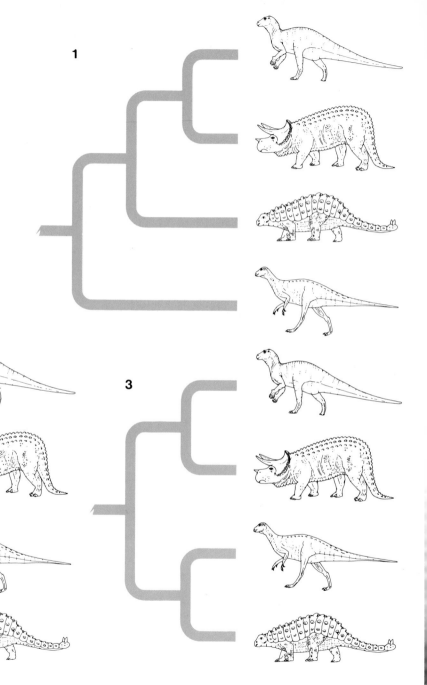

1

2

3

Do you agree that cladogram **2** is most likely to be correct?

Saurischian dinosaurs – how are they related to each other?

We know of no unique homology that the saurischians share, so we cannot call them a clade.
But we can look for homologies to test the relationships between individual saurischians.

Two of these saurischians share homologies not shared by the third.

Can you spot them?

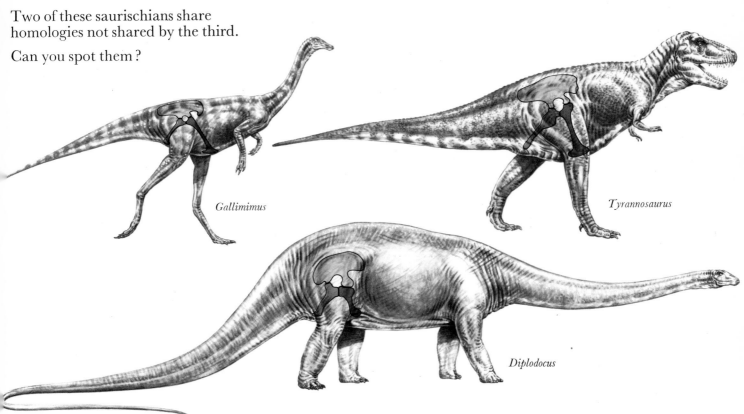

Gallimimus

Tyrannosaurus

Diplodocus

Tyrannosaurus and *Gallimimus* have similar shaped pubic (red) bones, and they walk on two legs.
These homologies are not shared by *Diplodocus*, so we think that *Tyrannosaurus* and *Gallimimus* are the two most closely related.

Dinosaurs and living archosaurs

This cladogram should remind you how we think dinosaurs are related to living archosaurs.

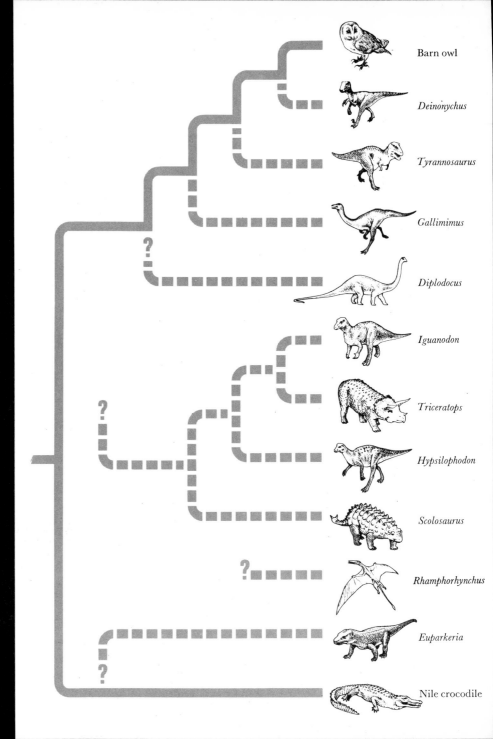

Barn owl

Deinonychus

Tyrannosaurus

Gallimimus

Diplodocus

Iguanodon

Triceratops

Hypsilophodon

Scolosaurus

Rhamphorhynchus

Euparkeria

Nile crocodile

In the next chapter you can test these suggestions by looking for homologies.

Chapter 6

Birds – the dinosaurs' closest living relatives

The cladogram on the opposite page suggests that birds are the living archosaurs most closely related to dinosaurs.

Before we test our ideas by looking for homologies, we must be sure that birds form a clade.

Can you think of a homology unique to birds?

All birds have feathers, and most birds have a 'wishbone'.* Both these homologies are unique to birds, so we assume that the common ancestor of birds was not shared by any other animal. Birds **do** form a clade.

*The 'wishbone' is small in emus and absent in other flightless birds, such as ostriches and moas.

Common emu
Dromaius novaehollandiae

wishbone

Rook
Corvus frugilegus

wishbone

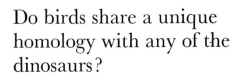

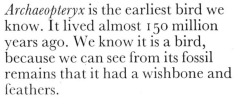

Archaeopteryx is the earliest bird we know. It lived almost 150 million years ago. We know it is a bird, because we can see from its fossil remains that it had a wishbone and feathers.

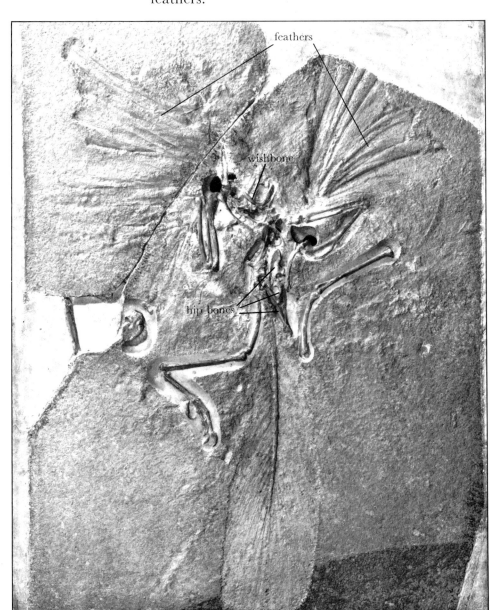

feathers

wishbone

hip bones

Do birds share a unique homology with any of the dinosaurs?

We know of no unique homology shared by dinosaurs and living birds. But living birds have lost a number of homologies shared by all other archosaurs. For example, they have no claws on the fingers, no teeth and no bony tail. We think that they lost these characteristics as they became adapted for flight. Since birds have evolved independently from other archosaurs for millions of years, it seems likely that they could have lost a unique homology they once shared with dinosaurs.

So let's look at an early fossil bird and see if it shares a unique homology with any of the dinosaurs.

Does *Archaeopteryx* share a unique homology with any of the dinosaurs?

First let's compare *Archaeopteryx* with ornithischian dinosaurs.

They have similar hips – both have a backward-pointing pubic (red) bone. But we think their hips are similar because they have similar functions, and not because they were inherited from a common ancestor.

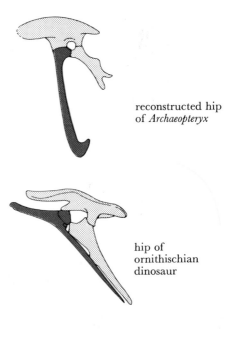

reconstructed hip of *Archaeopteryx*

hip of ornithischian dinosaur

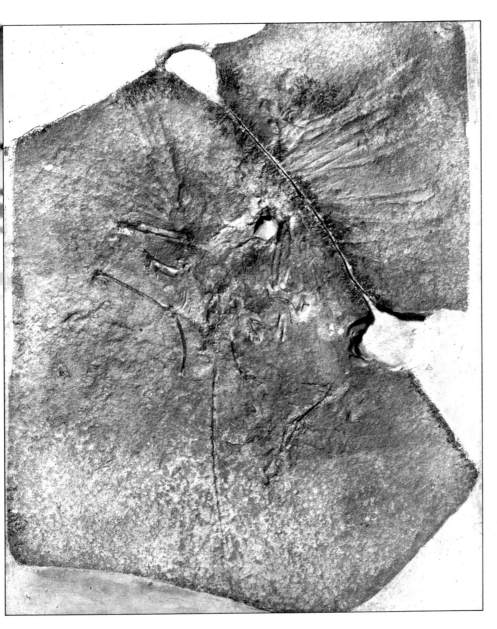

The first *Archaeopteryx* to be discovered – found in 1851 inside a limestone slab in a German quarry. The photographs show the two halves of the slab after it had been split open.

51

Now let's compare *Archaeopteryx* with some individual saurischian dinosaurs.*

Can you see that *Archaeopteryx* and *Deinonychus* have similar wrists? This homology is not shared by pterosaurs, thecodontians, crocodiles or any other animals. So we think that, of these animals, *Deinonychus* is most closely related to *Archaeopteryx* – and therefore to modern birds.

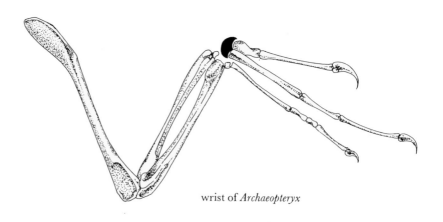

wrist of *Archaeopteryx*

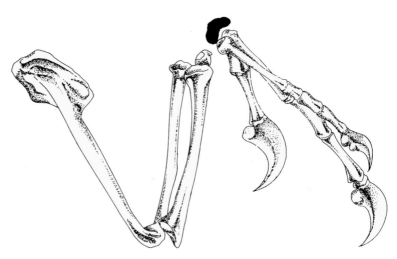

wrist of *Deinonychus*, a saurischian dinosaur

*Because saurischians do not form a clade, we must compare *Archaeopteryx* with **individual** saurischians.

Can you construct a cladogram to show how these four saurischian dinosaurs are related to each other and to birds?

The diagrams illustrate a number of homologies that are shared by just some of the animals. Use the homologies to work out how the animals are related, and construct a cladogram to represent your ideas.

		Tyrannosaurus	Deinonychus	Gallimimus	Diplodocus	Archaeopteryx
A 'foot' on one of the hip bones . . .						
A pair of bones shaped like this in the roof of the mouth . . .						
A wrist bone shaped like this . . .						
Only two fingers on each hand . . .						
A hip shaped like this . . .						

Cladogram of saurischian dinosaurs and birds.

Do you agree that this is how these saurischian dinosaurs are related to each other and to birds?

How likely do you think it is that birds are the closest living relatives of the dinosaurs?

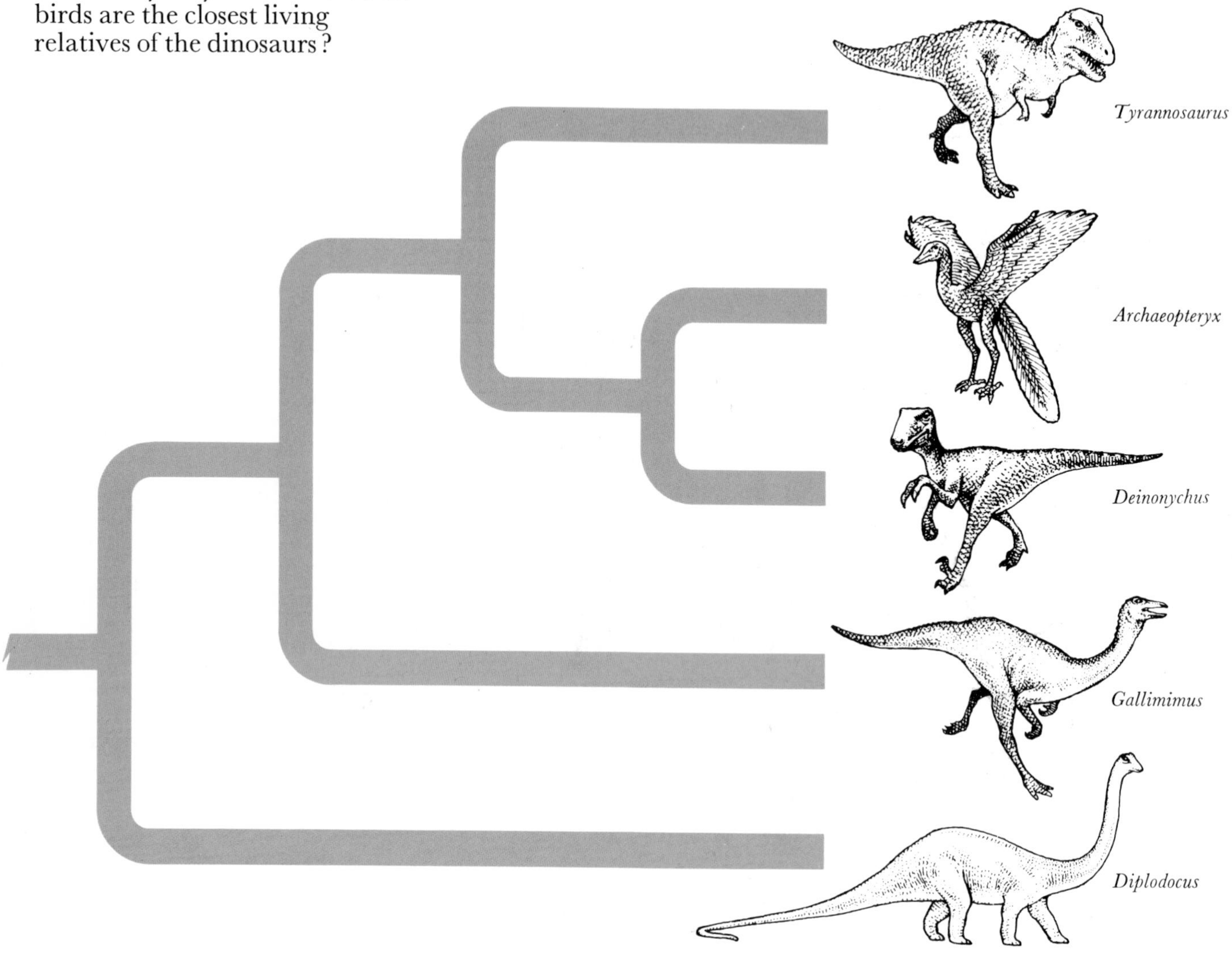

Tyrannosaurus

Archaeopteryx

Deinonychus

Gallimimus

Diplodocus

The evidence presented in this book suggests that birds are the closest living relatives of the dinosaurs. But some scientists think that birds and crocodiles are more closely related to each other than either of them is to dinosaurs. As research continues, there may be new evidence to suggest which idea is more likely to be correct.

We can never go back in time, so we can never know about all the animals that have lived in the past. But by working out the relationships between the animals we do know and then testing our ideas, we can begin to understand what has happened during the history of life.

Fossilized footprint of a dinosaur from a Purbeck Stone quarry near Swanage in Dorset. About 135 million years old.

The fossil animals featured in this book

Scolosaurus

Scolosaurus is an ornithischian dinosaur. In life, it was covered in an armour of thick bony plates, set close together in its leathery skin. The plates on its back had spikes and there were two very large spikes on the tip of its tail. Its armour would have protected it against flesh-eating dinosaurs such as *Tyrannosaurus*.

Scolosaurus probably fed on soft plant material – it had only one set of weak teeth.

Length about 5 metres
Age about 70 million years
Period Upper Cretaceous

	MILLIONS OF YEARS AGO
CAENOZOIC	0
UPPER CRETACEOUS	65
LOWER CRETACEOUS	136
UPPER JURASSIC	
LOWER JURASSIC	195
UPPER TRIASSIC	225
LOWER TRIASSIC	
UPPER PERMIAN	
LOWER PERMIAN	280
AGE OF THE EARTH	4600

Triceratops

Triceratops is a rhinoceros-like ornithischian dinosaur. It probably roamed in large herds, grazing the vegetation. It would have used its horny beak to cut up plants, and then chewed them with its rows of grinding back teeth. Its horns and the large bony frill over its neck would have protected it against flesh-eating dinosaurs such as *Tyrannosaurus*.

Length about 6 metres
Age about 70 million years
Period Upper Cretaceous

	MILLIONS OF YEARS AGO
CAENOZOIC	0
	65
UPPER CRETACEOUS	
LOWER CRETACEOUS	136
UPPER JURASSIC	
LOWER JURASSIC	195
UPPER TRIASSIC	
LOWER TRIASSIC	225
UPPER PERMIAN	
LOWER PERMIAN	280
AGE OF THE EARTH	4600

Iguanodon

Iguanodon is an ornithischian dinosaur. Its fossilized footprints indicate that it normally walked on two legs. But it has hoof-like bones on its hands which suggest that it sometimes walked on all fours. The 'thumb-spikes' on its hands could have been used for defence.

Iguanodon fed on plants. At the front of its mouth it had a horny beak instead of teeth. It would have used its beak to cut up the plants and then chewed them with the grinding teeth at the back of its mouth. It had several rows of teeth so, as old teeth wore out, they would have been replaced by new ones.

Length 4-9 metres
Age about 115 million years
Period Lower Cretaceous

	MILLIONS OF YEARS AGO
CAENOZOIC	0
UPPER CRETACEOUS	65
LOWER CRETACEOUS	
UPPER JURASSIC	136
LOWER JURASSIC	195
UPPER TRIASSIC	
LOWER TRIASSIC	225
UPPER PERMIAN	
LOWER PERMIAN	280
AGE OF THE EARTH	4600

Hypsilophodon

Hypsilophodon was a fast-moving, plant-eating ornithischian dinosaur. We think it was able to run fast because the lower parts of its legs were much longer than the upper parts – just as in fast-moving animals alive today. *Hypsilophodon* had a stiffened tail that was probably held out straight behind to help it balance as it ran. Running away was probably its best defence against enemies, but it may also have been protected by the bony lumps in the skin of its back.

Length 1.5 metres
Age about 115 million years
Period Lower Cretaceous

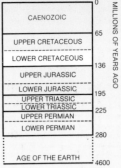

	MILLIONS OF YEARS AGO
CAENOZOIC	0
UPPER CRETACEOUS	65
LOWER CRETACEOUS	136
UPPER JURASSIC	
LOWER JURASSIC	195
UPPER TRIASSIC	
LOWER TRIASSIC	225
UPPER PERMIAN	
LOWER PERMIAN	280
AGE OF THE EARTH	4600

		MILLIONS OF YEARS AGO
		0
CAENOZOIC		
		65
UPPER CRETACEOUS		
LOWER CRETACEOUS		
		136
UPPER JURASSIC		
LOWER JURASSIC		
		195
UPPER TRIASSIC		
LOWER TRIASSIC		225
UPPER PERMIAN		
LOWER PERMIAN		
		280
AGE OF THE EARTH		
		4600

Diplodocus

Diplodocus is a saurischian dinosaur. It is one of the largest land animals that has ever lived. It was about 26 metres long and would have weighed about 10 tonnes.

Scientists used to think that *Diplodocus* lived in lakes or swamps, but we now think that this is unlikely for the following reasons:

● Most large animals that live in water (e.g. hippopotamuses) have a barrel-shaped body with a short neck and legs. *Diplodocus* is not this shape.

● *Diplodocus* has small feet in comparison with the size of its body, so it would have found it difficult to walk on soft, swampy ground without sinking in.

● The rocks where *Diplodocus* was found contain the remains of land plants and animals, and not swamp-living ones.

We now think *Diplodocus* probably lived on land.

Length about 26 metres
Age about 150 million years
Period Upper Jurassic

Gallimimus

Gallimimus was a slender, ostrich-like saurischian dinosaur with long powerful back legs for running fast. Some scientists think *Gallimimus* fed on small animals such as insects and lizards which it snapped up in its horny beak. Others suggest that it fed on plants or possibly even on the eggs of other dinosaurs.

Its three-fingered hands could have been used to tear up prey, but it seems just as likely that they were used for digging or searching amongst the vegetation for food.

Length 4 metres
Age about 70 million years
Period Upper Cretaceous

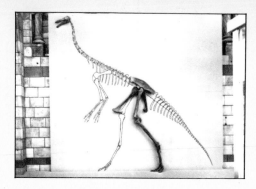

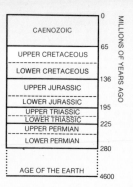

	MILLIONS OF YEARS AGO
CAENOZOIC	0
UPPER CRETACEOUS	65
LOWER CRETACEOUS	136
UPPER JURASSIC	
LOWER JURASSIC	195
UPPER TRIASSIC	
LOWER TRIASSIC	225
UPPER PERMIAN	
LOWER PERMIAN	280
AGE OF THE EARTH	4600

	MILLIONS OF YEARS AGO
CAENOZOIC	0
	65
UPPER CRETACEOUS	
LOWER CRETACEOUS	136
UPPER JURASSIC	
LOWER JURASSIC	195
UPPER TRIASSIC	
LOWER TRIASSIC	225
UPPER PERMIAN	
LOWER PERMIAN	280
AGE OF THE EARTH	4600

Tyrannosaurus

Tyrannosaurus is a saurischian dinosaur. It is the largest known flesh-eating land animal that has ever lived. It was 12 metres long, 5 metres high and would have weighed up to 7 tonnes. Its long teeth had sharp jagged edges for tearing flesh.

Tyrannosaurus had large powerful back legs that were probably used to grip its prey. It is difficult to imagine what its tiny front limbs were used for – they were so small they would not even have reached its mouth.

Length 12 metres
Age about 70 million years
Period Upper Cretaceous

Ichthyosaurus

Ichthyosaurus is not a dinosaur. It is an ichthyosaur. You can see from its shape that it is suited to life in water – it is streamlined, had fin-like paddles instead of legs, a tail fin and a fin on its back.

Ichthyosaurus lived all its life in the sea and probably fed on fish or shellfish (ammonites), which it caught with its sharp teeth. It had no gills so it could not take in oxygen from the water and, like dolphins and porpoises, must have come to the surface to breathe.

Ichthyosaurus did not even come on land to lay its eggs – it gave birth to live young in the water.

Length about 1.5 metres
Age about 180 million years
Period Lower Jurassic

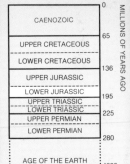

	MILLIONS OF YEARS AGO
CAENOZOIC	0
UPPER CRETACEOUS	65
LOWER CRETACEOUS	136
UPPER JURASSIC	
LOWER JURASSIC	195
UPPER TRIASSIC	225
LOWER TRIASSIC	
UPPER PERMIAN	
LOWER PERMIAN	280
AGE OF THE EARTH	4600

Plesiosaurus

Plesiosaurus is not a dinosaur. It is a plesiosaur. It lived in the sea and would have used its paddle-like legs to swim quickly through the water. It probably fed on fish which it snapped up with its sharp pointed teeth.

We do not know whether plesiosaurs came ashore to lay eggs or whether, like ichthyosaurs, they gave birth to live young in the water. They may have been able to move about on land, but only in a clumsy way like seals.

Length about 2 metres
Age about 180 million years
Period Lower Jurassic

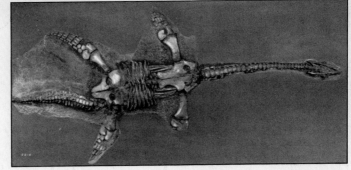

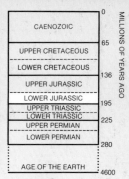

	MILLIONS OF YEARS AGO
	0
CAENOZOIC	
	65
UPPER CRETACEOUS	
LOWER CRETACEOUS	
	136
UPPER JURASSIC	
LOWER JURASSIC	
	195
UPPER TRIASSIC	
LOWER TRIASSIC	
	225
UPPER PERMIAN	
LOWER PERMIAN	
	280
AGE OF THE EARTH	
	4600

Pteranodon

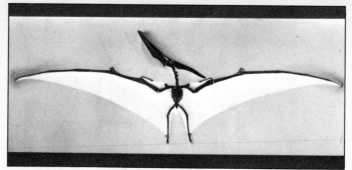

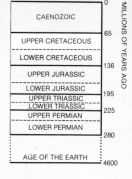

		MILLIONS OF YEARS AGO
CAENOZOIC		0
UPPER CRETACEOUS		65
LOWER CRETACEOUS		
UPPER JURASSIC		136
LOWER JURASSIC		195
UPPER TRIASSIC		225
LOWER TRIASSIC		
UPPER PERMIAN		
LOWER PERMIAN		280
AGE OF THE EARTH		4600

Pteranodon is not a dinosaur. It is a pterosaur. It is one of the largest known flying animals that has ever lived – it had a wingspan of 7 metres. But because it had hollow bones, *Pteranodon* was very light – adults weighed only about 16.6 kilogrammes (about the same weight as a large turkey).

Pteranodon probably flew by gliding with its large wings stretched out. It would have been very clumsy on the ground because it had weak legs and could not fold its wings completely, as bats and birds can.

Some scientists have suggested that *Pteranodon* lived on cliffs where it could take off into rising air currents. Its bones have been found in rocks formed from sea sediments, so it probably glided across the sea and swooped down to catch fish in its long toothless beak.

Wingspan up to 7 metres
Age about 80 million years
Period Upper Cretaceous

Archaeopteryx

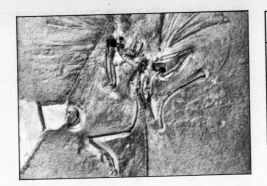

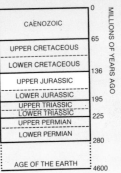

	Millions of years ago
CAENOZOIC	0
UPPER CRETACEOUS	65
LOWER CRETACEOUS	136
UPPER JURASSIC	
LOWER JURASSIC	195
UPPER TRIASSIC	
LOWER TRIASSIC	225
UPPER PERMIAN	
LOWER PERMIAN	280
AGE OF THE EARTH	4600

Archaeopteryx is the earliest known bird. We know that it is a bird because we can see from its fossil remains that it had a wishbone and feathers.

There are two ideas about the way *Archaeopteryx* lived. The traditional idea is that it lived in trees, and glided or flew from branch to branch. The other idea is that *Archaeopteryx* lived on the ground, and used to run along with its feathery 'arms' outstretched – perhaps to trap insects. As it ran along like this, it might have been able to lift off from the ground and fly.

But *Archaeopteryx* may not have been able to fly like a modern bird. When modern birds flap their wings, they use very large wing muscles that attach to a keel-shaped breastbone. *Archaeopteryx* does not have a breastbone like this, and its wing muscles may have been quite small. So, many scientists think that *Archaeopteryx* could only fly by gliding through the air. Some scientists now disagree with this idea; they think that *Archaeopteryx* was able to fly by flapping its wings.

Wingspan about 0.5 metres
Age about 147 million years
Period Upper Jurassic

Glossary

Amnion: a membrane surrounding the embryo of certain animals. It contains a fluid that keeps the embryo wet and cushions it against bumps and knocks.

Amniotes: a clade that includes birds, crocodiles, lizards and snakes, turtles and tortoises, mammals, dinosaurs, ichthyosaurs, plesiosaurs, pterosaurs and thecodontians. The unique homology that these animals share is an amnion.

Archosaurs: a clade that includes birds, crocodiles, dinosaurs, pterosaurs and thecodontians. The unique homology that these animals share is a particular hole in the skull in front of the eye.

Clade: a group of living things that includes all the descendants of a particular common ancestor.

Cladogram: a branching diagram that represents the way we think living things are related. On a cladogram, clades can be picked out.

Common ancestor: an ancestor shared by two or more living things.

Fossil: the remains of a living thing, or direct evidence of its presence, preserved in rocks. Usually only hard parts such as bones, teeth and shells are preserved.

Homology: a characteristic that is similar in two or more living things because it was inherited from their common ancestor.

Ornithischian dinosaurs: one of the two types of dinosaur. They are distinguished by having a hip shaped like this.

The group includes *Scolosaurus, Triceratops, Iguanodon* and *Hypsilophodon*. It is a clade because ornithischians share a unique homology – they have a predentary bone in the lower jaw (see page 42).

Saurischian dinosaurs: one of the two types of dinosaur. They are distinguished by having a hip shaped like this.

The group includes *Diplodocus, Tyrannosaurus, Gallimimus, Deinonychus*. We know of no unique homology that saurischians share, so we cannot call the group a clade.

Further reading

... about dinosaurs

Dinosaurs by Anne McCord, *The Children's Prehistory* series, Usborne 1977. For younger children.

A New Look at the Dinosaurs by Alan Charig, Heinemann/British Museum (Natural History) 1979. For older children and adults. An up-to-date account illustrated with many new and scientifically accurate drawings. Specially recommended.

The Evolution and Ecology of Dinosaurs by L. B. Halstead, Peter Lowe 1975. Also deals with ichthyosaurs, plesiosaurs, crocodiles and pterosaurs.

A Natural History of Dinosaurs by R. T. J. Moody, Hamlyn 1977.

The World of Dinosaurs by M. Tweedie, Weidenfeld & Nicolson 1977. Also deals with ichthyosaurs, plesiosaurs, crocodiles and pterosaurs.

Men and Dinosaurs: The search in field and laboratory by E. H. Colbert, Evans 1968. The history of dinosaur collecting in North America and Europe.

The Age of Reptiles by E. H. Colbert, Weidenfeld & Nicolson 1965.

Life before Man, Z. V. Spinar and Z. Burian, Thames & Hudson 1972. Deals with fossil amphibians, 'reptiles', birds and mammals.

... about fossil birds

Fossil Birds W. E. Swinton, British Museum (Natural History), 3rd edition 1975.

... about evolution

Evolution by Colin Patterson, British Museum (Natural History) 1978. An introduction to all aspects of evolutionary theory, written especially for those with little or no knowledge of biology.

... about living reptiles

The Life of Reptiles by A. d'A. Bellairs, 2 volumes, The Weidenfeld & Nicolson Natural History Library 1969. Also deals with fossil reptiles.

Reptiles by A. d'A. Bellairs and J. Attridge, Hutchinson University Library 1975. Also deals with fossil reptiles.

The World of Amphibians and Reptiles, Sampson Low Guides, 1978.

Living Reptiles of the World by K. P. Schmidt and R. F. Inger, Hamish Hamilton 1957.

The Reptiles by A. Carr, Time Life International 1964.

Acknowledgements for photographs
7: purple emperor butterfly, W. H. D. Wince/ Bruce Coleman Ltd; praying mantis, Rocco Longo/Bruce Coleman Ltd; dahlia sea anemone, David George; euglena, Oxford Scientific films; teal, K. W. Fink/Ardea, London; star sea squirt, David George; bird's nest fungus, S. C. Porter/ Bruce Coleman Ltd. 12: wild sheep, Hans Reinhard/Bruce Coleman Ltd; Madagascar flying fox, Heather Angel. 15: bluebottle, Stephen Dalton/Bruce Coleman Ltd; southern right whale, J. Bartlett/Bruce Coleman Ltd; fruit bat, Heather Angel. 17: southern right whale, Francisco Erize/Bruce Coleman Ltd. 18: dolphin, Heather Angel; salmon, Ian Beames/Ardea, London. 27: hatching smooth green snakes, Donald D. Burgess/Ardea, London. 28: frogs and spawn, Frank Greenaway.

Answer to page 7
2, 4, 5 and 6 are animals.
1 *Euglena* species
2 Star sea squirt *Botryllus schlosseri*
3 Bird's nest fungus *Crucibulum vulgare*
4 Dahlia sea anemone *Tealia felina*
5 Teal *Anas crecca*
6 Laboratory stick insect *Carausius morosus*

Index

Published by the British Museum (Natural History), London and the Syndics of the Cambridge University Press
The Pitt Building, Trumpington Street, Cambridge CB2 1RP
Bentley House, 200 Euston Road, London NW1 2DB
32 East 57th Street, New York, NY 10022, USA
296 Beaconsfield Parade, Middle Park, Melbourne 3206, Australia

© Trustees of the British Museum (Natural History) 1979

First Published 1979

Library of Congress Cataloging in Publication Data

British Museum (Natural History)
Dinosaurs and their living relatives.
Includes index.
  1. Dinosauria.   2. Birds.   3. Crocodiles.
I. Title.
QE862.D5B68 1979   568'.19   79–14504

ISBN 0 521 22887 5
ISBN 0 521 29698 6 pbk.

Printed in Great Britain by W. S. Cowell Ltd, Ipswich